Number play

David Kirkby

Heinemann

00111

First published in Great Britain by Heinemann Library
an imprint of Heinemann Publishers (Oxford) Ltd
Halley Court, Jordan Hill, Oxford OX2 8EJ

MADRID ATHENS PARIS
FLORENCE PRAGUE WARSAW
PORTSMOUTH NH CHICAGO SAO PAULO
SINGAPORE TOKYO MELBOURNE AUCKLAND
IBADAN GABERONE JOHANNESBURG

© Heinemann Library 1995

Designed by The Pinpoint Design Company
Printed in Hong Kong by Wing King Tong Co. Ltd.

00
10 9 8 7 6 5 4 3 2

ISBN 0431 07932 3

British Library Cataloguing in Publication Data available on
request from the British Library.

Acknowledgements
The Publishers would like to thank the following
for the kind loan of equipment and materials
used in this book: Boswells, Oxford; The Early Learning
Centre; Lewis', Oxford; W. H. Smith; N. E. S. Arnold.
Special thanks to the children of St Francis C.E. First School

Photography: Chris Honeywell, Oxford

Cover photograph: Chris Honeywell, Oxford

There are 4 sets of 13 cards in a pack of playing cards. The 4 sets are hearts, clubs, diamonds and spades.

Here is a pack of playing cards.

Which card comes first?

To do:
Shuffle a pack of cards.
See how quickly you
can lay them out
in the right order.

In some card games you must collect sets of cards. To win this game you must collect 4 cards of the same number or 4 cards of the same kind in order.

Ian needs one more card to win. The 5 of spades.

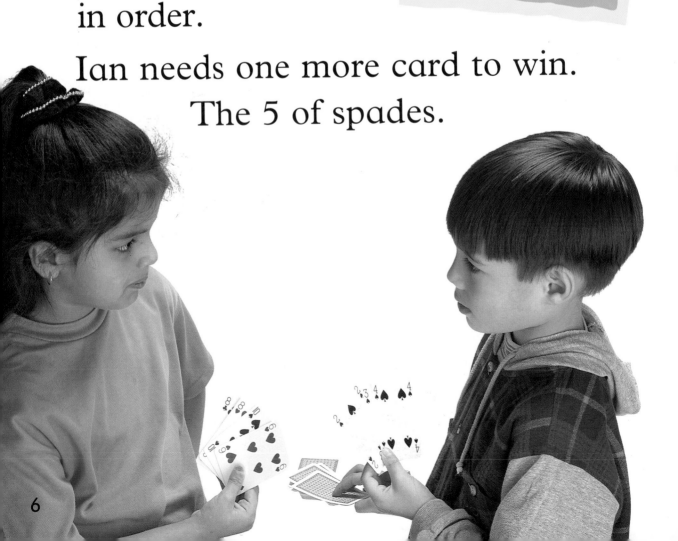

What card does Nadia need to win?

To do:
What pattern do
these numbers have?

You can play games with number patterns.

You can play hidden number games with lots of different things.

1	2	3	4	5
6	7	⬤	9	10
11	12	13	14	⬤
16	⬤	18	19	20
21	22	23	24	25

Which numbers are hidden?

To do:
Copy the number square. Use a ruler. Fill in all the missing numbers.

On a dice the number
of spots on the top
and the number on
the bottom always
add up to 7.

When 5 spots are on top,
2 spots are on the bottom.

When 4 spots are on the bottom,
3 spots are on top.

How many spots are on the bottom of these dice?

To do:
Throw a dice 10 times.
Guess how many
spots on the bottom
each time.

In race games the first person to get to FINISH, wins. You throw a dice to see how many spaces you can move.

If your counter is on a square at the bottom of a ladder, you go up to the top of it. If your counter is on a snake's head, you slide all the way down its tail.

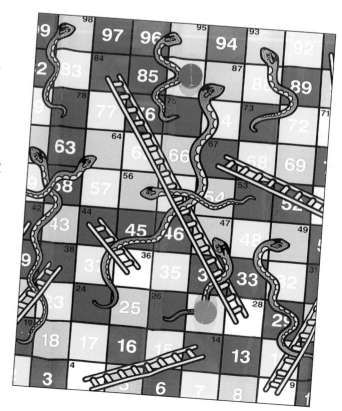

Which colour is winning?
What number does red need
to throw to win?

To do:
Make up your own
race game with ways to
go up and down.

Dominoes is a matching game. You have to match the pictures or the number of spots.

These dominoes have been matched.

Who can go next?

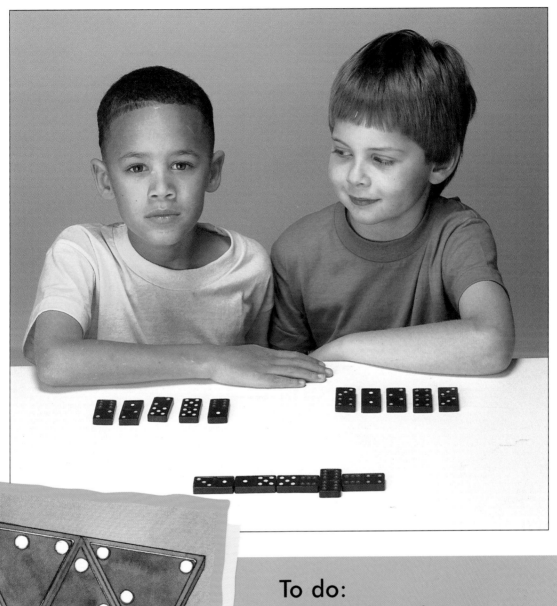

To do:
What other shapes
make good dominoes?
Try triangles, squares
and circles.

In lots of games you
must spot things
that are the same.

Monica is trying
to match the pairs.

How many pairs are there?

To do:
How many pairs
of coins are there in
this picture?

Some matching games are more difficult. In this game, pairs add up to the same total.

These pairs add up to 8.

18

These pairs must add to up to 10.
Which are the pairs?

To do:
What are the pairs
that add up to 6?

19

In some games
you need to use
your memory.

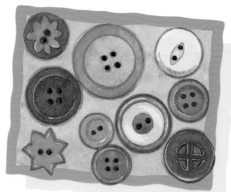

Look at the buttons very
carefully and try to remember
how many there are,
and what colours they are.

Look at these things carefully.
Now close the book and see
how many things you can
remember.

To do:
Collect 5 things yourself.
Now play the memory game.

These numbers make a pattern.
The even numbers are red.
The odd numbers are yellow.

Judy is finding all the odd numbers. The next number will be 9.

Which cards come next in this pattern?

To do:
What are the next 2 cards
in this counting down
pattern?

answers

index